Topsy a

AT THE HAI

Jean and Gareth Adamson

BLACKIE : GLASGOW AND LONDON

ISBN 0 216 90691 1 (hardback)
ISBN 0 216 90690 3 (paperback)

Blackie and Son Limited
Bishopbriggs, Glasgow G64 2NZ
Furnival House, 14-18 High Holborn
London WC1V 6BX

Printed in Great Britain by
Thomson Litho Ltd, East Kilbride, Scotland

One Saturday morning, Topsy and Tim's
Mummy went to get her hair done.
Topsy and Tim stayed home to
help Dad mend the kitchen tap.

Mending the tap was a wet, squirty job. It was difficult too, and made Dad a bit cross. Topsy and Tim began to wish they had gone with Mummy to the hairdresser.

"May we play hairdressers, Dad?" asked Topsy.

"Yes," said Dad. "Blow this wrench!"

Mummy's dressing-table had a big looking-glass, like the hairdressing salon. Topsy found the scissors and Tim found a comb.
Topsy cut Tim's hair.
Tim cut Topsy's hair.

"We don't look like
us any more," said Topsy.
"That looking-glass
must be wrong,"
said Tim.

Mummy came home, looking neat
and smelling sweet.
"Oh, Topsy and Tim!" said Mummy.
"What have you done!"
Mummy told Topsy and Tim they were
naughty to play with sharp scissors.
"It's lucky you didn't hurt yourselves,"
she said, "but . . . oh, just look
at you!"
Then Mummy could not help laughing.

Mummy took Topsy and Tim
to the hairdresser.
"They can't put your hair back,"
she said, "but they will make
what is left look tidy."
Topsy and Tim wore their big
sou'wester hats, although
it was a fine day.

Barber
FRUIT

Pauline, the hairdresser, made
Topsy's hair look as though
it was meant to be short.
"Now you, Tim," she said. But Tim
would not let Pauline trim his hair.

He bellowed when Mummy tried
to sit him in the chair.
"Well, we are ladies' hairdressers,"
said Pauline, "and Tim is a little
man. Why not take him to
his Dad's barber?"

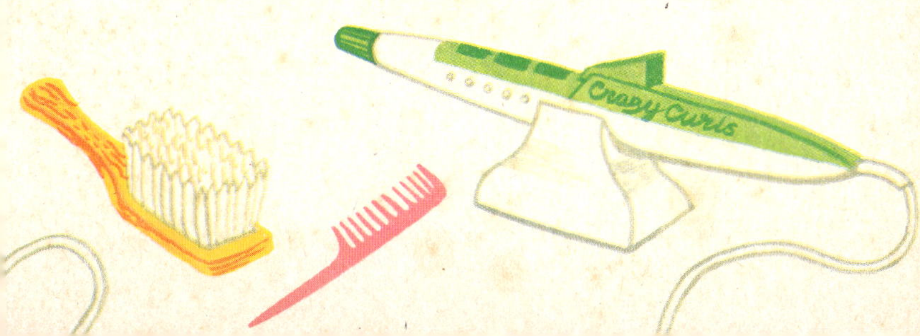

Topsy and Tim and Mummy had a long wait in the barber's shop. When their turn came, Mr Randall, the barber, said, "Which little boy shall I do first?"
He had thought Topsy was a boy with her hair so short.

Tim did not want his hair trimmed by Mr Randall either. Mr Randall showed him how the barber's chair went up and down.
They were soon good friends.
Mr Randall cut Tim's hair very short indeed.

"Do you want dressing?" he asked.
"I got dressed this morning," said Tim.
But the dressing Mr Randall meant was a sort of hair cream in a big bottle. He rubbed some into Tim's last bit of hair and made it glossy.

"I want dressing," said Topsy.
"Why not?" said Mr Randall.
He hoisted Topsy into the chair
and made her hair glossy too.

As soon as they got home, Topsy and Tim ran upstairs to Mummy's dressing-table, to look at themselves.

After a while
they heard Dad call,
"Topsy and Tim,
come and show me
your new haircuts."

Dad got a surprise when Topsy and Tim came downstairs. They had so much hair they could hardly see where they were going.

"Mr Randall's dressing must have made Topsy and Tim's hair grow again," said Mummy.
But really Topsy was wearing Mummy's old dress-wig. Tim's new hair was Mummy's furry winter hat.

Topsy and Tim's real hair
took a long time to grow again,
but it did at last.
Here they are, to prove it.